Song Birds

With an introduction by D.H.S. Risdon

CRESCENT BOOKS

Contents

3   Introduction

4   Why birds sing

6   Aviculture

9   Song bird families

12 Recording the song of birds

14 Song birds: habitats and nests

14 Bibliography

16 Index

17 Illustrations

Picture credits: Bruce Coleman 1, 4, 7, 8, 9, 54; Eric Hosking 2, 3, 5, 6, 11, 13, 14, 15, 19, 20, 24, 25, 27, 28, 29, 32, 34, 36, 38, 41–44, 46, 48–50, 52, 55, 65, 66; R. Longo 10, 12, 16–18, 21, 23, 30, 31, 33, 37, 39, 45, 47, 56–59, 62, 64, 67, 70, 72–74, 76, 77; C. Bevilacqua 22, 53, 61, 63, 69, 71, 75; A. Margiocco 26, 40, 51; M. Rota 35, 60, 68

Adapted from the Italian of Armando Grasselli
Edited by D. H. S. Risdon

Copyright © 1968 by Istituto Geografico de Agostini, SpA, Novara
English edition © 1974 by Orbis Publishing Limited, London
Library of Congress Catalog card number 73-91935
All rights reserved
This edition is published by Crescent Books
a division of Crown Publishers Inc.
by arrangement with Orbis Publishing Limited
Printed in Italy by IGDA, Novara

Introduction

Writing a book on song birds, absorbing though it may be, has the great drawback that the very essence of it cannot be put into words. The song of birds is, of course, impossible to translate into human terms.

It is hoped that with the help of this book and its illustrations the reader will have the pleasure of first discovering and then getting to know all the birds found in his own and other countries. They form part of that attractive side of nature which we all enjoy. It is true to say that the song of birds and the colour of flowers are numbered among those things in life which can induce happiness both in children and in old people, both in the amateur and the professional. Collected in this volume are pictures of birds from all over the world; many of them can be looked for and found in our woods and grassland or in the birdhouses of aviculturists. In the introduction the reader will find advice regarding the care and successful breeding of those species suitable for aviculture, and a section on the recording of bird song, an activity which both adds to the interest of bird watching and allows us to recapture in our homes some of the pleasure that bird song gives us.

Following the moult all birds cease their song for a long or short period. Nevertheless for no part of the year is bird song entirely absent, and for anyone new to the study of birds it is suggested that they keep a record of the time of year they first hear the song of the various species. Even in January, when there is less bird song than in any other month, the robin sings – and he continues to do so to the beginning of July when his moult starts. Mistle thrushes and starlings, too, can be heard in the early months of the year, and in March the blackbird, one of the most endearing of song birds, starts his song. Then for the next three months he delights us until his moulting period at the end of June. Birds that migrate for the winter re-appear in spring, the first in March usually being the chiffchaff. Other warblers follow in April. If we are lucky we hear in May perhaps the most beautiful example of bird song – the Nightingale's clear, sustained notes. This heralds the high summer of bird song which enriches our lives. The memorable song of this bird is recorded by Wordsworth in the lines:

'O Nightingale thou surely art
A creature of a fiery heart:–
These notes of thine – they pierce and pierce;
Tumultuous harmony and fierce!
Thou sing'st as if the God of wine
Had helped thee to a Valentine.'

Why birds sing

It is true to say that all birds sing in the sense that they utter certain sounds or notes, although there are some birds that are remarkably silent, such as the stork. The concept we have of bird song is a very personal one; we love those birds and those songs which most satisfy our ear. No one would ever dream of keeping a quail in order to listen to its characteristic call. Accordingly one must, in practice, exclude all those birds which do not belong to the passerine family of perching song birds (Passeriformes). The sounds uttered by non-passerines are completely unmusical to the human ear.

Common hangnest

As for the reason why birds sing, experts have given many different interpretations. From the scientific point of view the true song is closely connected to sexual behaviour; it is in fact related to all those display activities that the male must carry out in order to procure himself a mate, keep her and defend his own nest-building territory. However, in spite of these scientific theories, it does seem that birds also sing from a sense of well-being, as one can easily tell by watching and listening to them. In addition one can observe young males that try out and endlessly rehearse their songs while still in the first year of life and when no immediate necessity urges them to do so. One is automatically reminded of the vocalizations of singers before dress rehearsal and, in fact the two are substantially very similar. In this connection it is worth noting that birds inherit the call notes of their species and use them instinctively even when born and bred in a cage, whereas the most gifted song birds have, in addition, to learn from their elders the more complicated passages of their repertoire.

It is known that various bird populations of the same species have different songs depending on their surroundings. This is comparable to the existence of dialects in humans. For the newcomer to the subject some basic facts are worth stating. Birds are enabled to sing because a particular physiological conformation of the trachea which, at the junction of the two bronchi, contains various membranes. In particular it contains a vertical reed and certain muscles which, by contracting, stretch the membranes and make them vibrate, thus modulating the sounds. This vocal apparatus is very well-developed in the Passeriformes, although it is more rudimentary in the other orders of birds. Cranes possess a peculiar trachea which allows them to utter certain resounding notes audible at a great distance. Pigeons, on the other hand, can inflate their crops with air, then expel it producing the very characteristic and well-known sound. Other sounds can be produced mechanically in various ways; by clattering of the beak, a method used by the stork, or, like the woodpecker, by drumming on tree-trunks. In this way certain birds manage to communicate among themselves just as some African tribes do on their drums. The 'drumming' of snipe in courtship flight is produced by the two outer feathers of the tail when held stiff at a certain angle. Humming birds can produce a loud click with their wings when warning a rival.

Song birds can be divided into three vocal categories: (a) the true singers, for example nightingales, thrushes, blackcaps; (b) birds which go on repeating a musical phrase, for example, chaffinches, Golden Orioles; and (c) birds which repeat notes with rhythmical variations, for example cuckoos, chiffchaffs. As already mentioned, the reason why a bird sings has, generally speaking, a sexual origin. That is to say the male establishes, by singing, his exclusive rights over a certain territory, thus warning all other males of his species to keep at a required distance.

The variations on this main theme are many. The song and call notes of each species also serve as a means of identification between individuals. Just as the song may warn off rival males, it undoubtedly helps to attract females to the male's territory. There are birds that abandon their nesting territory almost immediately after the breeding season. There are others that claim it all the year round, such as the robin, which cannot tolerate the presence of any individuals of its own species even during the winter. The general rule (with exceptions) indicates that the better songsters are quietly coloured. This is explained by the fact that a very conspicuous male uses his colour to attract the female, while the bird with dull, uniform plumage must,

Grey fantail

of necessity, possess a good voice in order to attract the opposite sex.

Forest birds that inhabit the thick, dark undergrowth have strong and clear voices so that they can be heard a long way off. Marshland birds display this characteristic too, witness the marsh warbler, rails and crakes. The male skylarks, that inhabit open spaces and grassland, sing because of their inconspicuous plumage, which, if it serves its purpose as protection against enemies, so closely resembles that of the female that it may well be an obstacle to the male in its search for a mate. In order to deliver their song, males usually chose a prominent place in full view. Blackbirds, song thrushes, starlings, redstarts, melodious warblers and mocking birds come into this category. True forest and marsh birds can deliver it also while hidden in the thick of the forest or undergrowth. Lastly, many birds deliver their song in flight.

The singing activities of birds are extensive and are linked to a whole range of behaviour patterns and display actions, in addition to the one already mentioned.

The utterance of songs or call-notes can be classified as follows:

Call-notes of contact for migratory flocks for birds in flight. This is a type of call-note that is used to hold the flock together during migratory flights. There often can be hundreds of individual birds on these migrations. This particular type of call-note is most in evidence with ducks and geese; these call to one another at regular intervals during flights. Many other species use this type of call-note in order to maintain contact, or even to ask for permission to land from a group of friends already on the ground.

Search for food. The call-notes made in the search for food are of a special kind; they are noticeable among members of the same species, such as gulls and crows, and also among members of different species which gather together in winter and always act collectively, without losing contact, thanks to these particular call-notes. This is the case, for example, with flocks of tits and finches that live in the woods in winter.

Alarm-notes. These are particular sounds uttered by birds at the sudden sight of an enemy, such as a falcon in flight, in order to warn their fellows. These sounds can also be of a defensive-offensive nature, as with the chattering of sparrows at the sight of a buzzard or the hubbub of tits when mobbing an owl.

Call-notes between parents and their young, and vice versa. These notes are uttered only during the breeding season and they are used mainly by the young calling for food or by the parent birds as a warning to their young to keep still when there is danger.

Courtship song. By this generic term is meant the true song, already dealt with at length. Courtship songs present varied characteristics: aggressiveness towards other males, a general defensive attitude by the bird to its own territory,

Jackdaw

or the means of letting the female know that her mate is near and is guarding the nest.

Adequate interpretation of bird sounds, outside the categories already mentioned, is an extremely arduous task. These sounds are composed of such a range of tones that only the most exhaustive study can interpret them. Modern methods of registration have verified that the same bird can produce hundreds of variations of one musical phrase. Today in North America a number of Institutes have imposing collections of bird song recordings, as has the BBC. In France and elsewhere gramophone records of bird songs are easily obtainable.

The interpretation of bird song today reveals certain systematic affinities which were not noticed by earlier researchers because they lacked the opportunity of comparing the various modulations of a song. It is possible now because we can make use of tape recordings, sonograms, (a sonogram is a graph of pitch, in kilocycles, against time, in seconds). According to their living habits, birds sing either during the day or night. Owls being night-birds call during the hours of darkness, while the blackbird and the chaffinch sing during the day. Generally speaking the majority of day-birds sing at dawn and dusk, and stay silent during early afternoon. Typically twilight-birds are the goatsucker and the nightingale. At night one can sometimes hear the song of birds which are normally day-birds, such as the nightingale, the woodlark and the Great Reed Warbler.

Northern willow tit

The duration of the season when bird song is heard is variable. Many birds sing all the year round; others do so only for a very short time. The skylark, for example, sings continually from March to July, while the blackbird begins to sing softly in February, reaches its peak in June, then gradually lowers its tone and is completely silent for the rest of the year. The robin sings almost all the year round, but reaches its highest peak in March and September and its lowest in July. The wren, which also sings all the year round, does so in a more uniform way.

The song of birds has from time immemorial caught the imagination of man and, above all, of poets and writers. In popular proverbs and folklore are many references to the songs of better-known birds, among which are the cuckoo, the mocking bird and the blackbird. A famous poem is that on the skylark by Shelley. Minor and more fragmentary references can be found in many other writers as far back as Roman times and in Homer and the Bible.

Birds and their songs have, at times, been interpreted, (sometimes tragically for the bird) as bearers of ill-fortune. These birds have the misfortune of sounding mournful to human ears. This happens with some of the nocturnal birds of prey, which, due to their appearance and their strange call-notes, are regarded as bad omens.

Many birds have been named after the sound they utter, as, for example, the Chiffchaff that is called Zilpzalp in German and Tjiftjaaf in Dutch, precisely because of the sound this bird utters most frequently. In Britain the cuckoo is the best known example.

Loddiges' racket-tailed hummingbird

Aviculture

Aviculture is the name for the care and breeding of birds in captivity, just as horticulture means the cultivation of plants.

While it is largely the activity of the private hobbyist who keeps birds for the sheer pleasure they bring, it is becoming more of an exact science. As more and more species are bred in confinement so does our knowledge of bird behaviour increase. There is no doubt that the serious bird breeder has much to contribute concerning the private lives of birds. After all how better to get to know birds than to live with them?

Marsh warbler

For those who wish to take up this most satisfying and fascinating of pursuits here are some guidelines. First you should have a natural inclination for animal husbandry. Birds are not toys to be put on one side when you are tired of them. The care of any livestock carries with it certain obligations. It is as well to realize at the outset that living things can be a tie requiring daily attention throughout the year.

On the other hand, birds make less demands on their owners than many other animals, except perhaps for the single specimen kept as a house pet with no other companionship than that of its owner. They live lives of their own which are very different from ours and it is a mistake to imagine them like ourselves.

Living quarters for birds should be light and airy. They are creatures of the light and fresh air. Dark stuffy quarters are bad for them.

Always provide plenty of space. Cages and aviaries should be as large as possible. The all-wire type of cage usually sold in pet shops is adequate for a pair or so of small birds, but in many ways the box type is better – that is one closed in all round except for the wire front. The box cage gives the occupant a greater sense of seclusion and is also free from draughts. Needless to say such cages must face the light.

Whatever type of cage is used, it should be fitted with a

drawer tray to facilitate clearing it out. The floor can be covered with sand for seed-eaters which will eat some of it as grit to aid digestion. Insect and fruit-eaters will not however need grit and their floor is better covered with sheets of newspaper cut or folded to the size of the drawer tray.

Perches should not be too thick. They should be thin enough for the bird to get its claws right round. Perches need regular cleaning and should be scrubbed in hot water now and again.

The best way of keeping birds is in an aviary – preferably an outdoor one. Contrary to popular belief even tropical birds will acclimatize to living outdoors in temperate countries, provided that the enclosure is properly built with

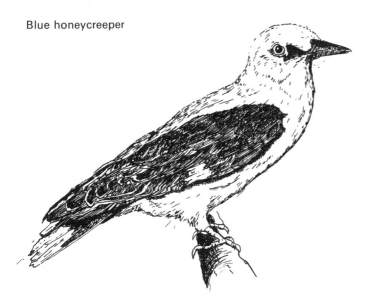

Blue honeycreeper

a snug well-lit shelter to which the birds can retire at night and in bad weather.

The shelter is the most important part of the aviary as it is here that the birds congregate, so the shelter really governs the number you can keep. It must admit plenty of daylight or the birds will not use it. This can be provided either through windows in the side or by using PVC corrugated sheeting on the roof.

The best and most comfortable perches should be under the shelter and at least some of the feeding should be done inside it. In fact if everything is done to encourage the birds to use it during the daytime they are more likely to roost in it at night.

Another advantage of persuading the birds to use the shelter at night is the prevention of night frights. Cats or owls on top of an aviary at night can cause havoc among birds roosting in the open. Car headlights will also cause them to panic in the dark and bang their heads with fatal results.

The open-air part of the enclosure can be of any size or shape you like and, if large enough, can be planted with grass and bushes so that it forms part of the garden. It is, however, a mistake to try and plant *small* aviaries. The constant nibbling of the birds and their droppings on the leaves will look unsightly and eventually cause the death of the plants.

Birds left in outside aviaries get the benefit of direct sunshine, fresh air and rain which will give their plumage a gloss seldom acquired indoors. Moreover, they are less trouble to look after.

Song birds fall into two classes, called by fanciers hardbills and softbills. Hardbills have beaks capable of cracking open seed, discarding the husk and eating only the kernel. Softbills eat insects and fruit. In other words their diet is of a soft nature. It is as well to remember, however, that both hardbills and softbills are to a certain extent omnivorous. Although the diet of seed-eaters may be mainly seed, they do eat insects and greenfood, especially when breeding.

Hardbills need grit to enable them to digest their food. Softbills do not. In fact grit could do them harm.

Seed-eaters are easier to manage as their food keeps fresh and all that has to be done is to blow off the empty seed husks every day. Softbills are provided with a ready-made mixture of biscuit meal, dried flies, ants' eggs and meat meal to which vitamins are added. This is sold already mixed in packets. They also require fresh fruit and a few mealworms, maggots or other live insects daily. The moist nature of their diet inevitably means greater attention to cleanliness.

Cuttlefish bone is liked by all song birds, particularly seed-eaters. It provides them with extra lime which is good for feather and bone formation.

All song birds love bathing. This is most important to keep their plumage in condition. Most of them will bathe daily if given the chance. Bathing facilities should be provided in the form of wide shallow dishes filled with *clean*

Golden oriole

water. The water must be clean or the birds will not bathe in it. The old-fashioned earthenware flower pot saucers are ideal bathing dishes if you can get them. If not, the modern plastic ones will do.

The breeding of birds is probably the most interesting side of aviculture. The domesticated species like Canaries, Zebra finches and Bengalese, and several of the Australian Grass finches are much the easiest to breed, but nowadays many other species are raised regularly in captivity. While the domesticated birds will breed fairly well in colonies, the best results are obtained when each pair are given separate quarters.

A mixed collection of unmated birds will usually agree if carefully selected, but once a pair starts to breed it becomes territorially-minded and will turn on its erstwhile companions. If several pairs get like this they will spend the time quarrelling and never get down to serious breeding. A lot depends on the size of the aviary. Obviously the larger the enclosure the more space the birds have to get away from each other.

If the aviary is naturally planted with thick bushes the birds will find some of their own nesting sites. Otherwise these have to be provided in the form of clumps of twiggy branches fixed up in secluded corners, wooden nest boxes with a hole in the side and similar boxes with half the front

Green catbird

open. Special nest pans made of earthenware or plastic can be bought as well as wicker nests and wooden nest boxes.

Sick birds should be treated by placing them in a very warm temperature. This should be at least 85° to 90°F. Special glass-fronted, electrically-heated hospital cages can be bought for this purpose. Failing this, an ordinary cage can be draped with a blanket except for the front which should be placed opposite the source of heat. Hang a thermometer in the back of the cage and adjust the distance away from the heat till the correct temperature is obtained.

It is also a good plan to add an antibiotic to the drinking water in case there is an infection, but this can only be obtained on veterinary prescription. Aureomycin is a good general purpose antibiotic.

Most of the birds illustrated in this book are suitable for aviculture, but it is as well to remember that many of them are protected in their native countries and it is illegal to catch them. Aviary-bred strains of quite a lot of species are, however, obtainable and these make the best subjects.

Canaries have been domesticated for hundreds of years, and now exist in various colours – green, cinnamon, yellow, white, grey and orange. Being used to man and his ways they are easy to keep and breed either in cages or aviaries and will rear several broods in a season. They also sing beautifully. In fact one variety – the Roller – is specially bred for the beauty of its song. The canary 'fancy' is a hobby in itself. Societies exist for owners of each breed.

The Zebra finch lives wild in Australia, but is now thoroughly domesticated and exists in many different colours. The original wild form is beautiful enough, but imitations have occurred over the years and have now been established in silver, fawn, white and pied – the latter being a broken colour pattern of dark colour and white. This pretty little bird is very prolific and will produce brood after brood all the year round if allowed to do so. All it requires is a nest box with a hole in the side which it will line with hay and feathers. It will breed in a cage or an aviary.

Many of the Australian Grassfinches have now been established as free breeding strains in European and American aviaries, such as the very beautiful Gouldian finch, the Long-tailed Grassfinch, the Cherry finch and the Diamond sparrow. They have the same requirements as the Zebra finch.

A little bird from Asia – the Bengalese – is another easy species to keep. In America it is called the Society finch because of its habit of crowding with all the other Bengalese finches into one box to sleep at night. It exists in several colour varieties such as chocolate, white and fawn. The pied form consists of a broken colour pattern in which one of the above colours is mixed with white. The Bengalese also requires the same treatment as the Zebra finch.

Of the birds illustrated in this book the following do well in captivity: starlings, thrushes, blackbirds, and the Pekin robin. These are all softbills and require insectile mixture as already described together with fruit and a few mealworms or maggots daily. They are hardy and present no problems to the aviculturist provided their quarters are kept very clean and they have ample opportunity for daily bathing. All the seed-eaters mentioned, particularly the finches and waxbills, are to be recommended.

Most of them will breed readily in aviaries if given the chance and the right surroundings.

European dipper

Song bird families

In this section mention will be made of the various families, which are justly called 'song bird families':

Larks *(Alaudidae)* This family includes such well-known names as skylarks, crested larks and calandra larks. They are typical meadow-birds which rarely perch on trees. Their plumage is inconspicuous, but their voices are strong and melodious, and audible at a great distance. Members of the family are distributed all over the world, from arctic tundra to barren deserts. Generally speaking, all skylarks sing in flight and also during seasons other than spring, the usual season for bird song. Their song has inspired poets like Shelley, Wordsworth and Tennyson. Larks eat all kinds of insects and seeds. Europeans so loved the skylark that they took it with them to many parts of the world where it has successfully established itself in places like New Zealand and the Hawaiian Islands. Most larks build open cup-shaped nests on the ground, well camouflaged among long grass.

Pipits and wagtails *(Motacillidae)* This family is made up of terrestrial – or ground-birds – that dwell primarily in moist grassland and along waterways. The best songsters of the family are those with the least conspicuous plumage. To this group belong about 54 different species found in many parts of the world. Pipits are mainly ground dwellers. They are dull coloured in varying shades of mottled brown and in some ways remind one of larks. The different species are much alike and are not easy to identify in the field. They build open cup-shaped nests on the ground concealed among long grass. Wagtails are more conspicuously coloured than pipits, their bright patterns of black and white or grey and yellow making them very noticeable. They tend to inhabit gardens and other man-made places which also brings them to our attention. Their most noticeable characteristic is their long tails which are constantly wagged up and down as the birds run over the ground. These dainty pretty birds are adept at catching insects on the wing. They nest in holes and crannies in old walls and buildings.

Black-and-white warbler

Orioles *(Oriolidae)* This is a small family of tropical or subtropical birds with arboreal habits. All orioles are brightly coloured and have a very characteristic song. Europe can claim only one of the 34 species that make up the family. The true orioles are confined to the Old World and are not related to the American orioles although they resemble them superficially. The word itself is derived from the Latin 'aureolus' meaning gold or yellow, so any black and yellow bird tends to be called an 'oriole' – an example of how vernacular names can cause confusion. These birds are about the size of starlings and have fine clear songs. The males have the bright yellow plumage. Hens are of a duller green hue. Their nests are slung like hammocks between two branches.

Tui or parson bird

Starlings *(Sturnidae)* A gregarious family of birds; usually with rather dark plumage carrying beautiful metallic hues. Starlings are active and talkative. Some members of the family are well-known in Africa because of their habit of searching for parasites on the backs of large mammals. The voice of the starling is not particularly melodious, but the bird can mimic other notes and phrases. The glossy starlings from Africa are some of the most brilliantly coloured birds in the world. Their plumage has a metallic sheen of green, blue and purple which varies according to the light and makes them look as if cast in burnished metal. Starlings are a most successful group of birds. They all come from Africa and Asia, but the common starling has unfortunately been introduced by man to many parts of the world where it has established itself and become a nuisance. In Eastern countries starlings are called mynahs and there are many different species. To this family belongs the Indian mynah *(Gracula religiosa)* which can copy the human voice. The bird is able to repeat words and some can learn whole sentences.

Thrushes, chats, wheatears, redstarts, robins *(Turdidae)* This large family is distributed throughout the world; to it belong about 300 species of fruit-eating and insect-eating birds. The habits of members of this family are varied and depend on environment. They can be found in forests as well as in deserts, on mountains as well as in marshes. Many of the *Turdidae* are strong migrants. Almost all are excellent songsters and some the very best in the world, for example the song thrush and the nightingale. In spring many of them are brightly coloured. The American Robin and the famous Bluebird are both members of this family. Thrushes build open cup-shaped nests in bushes and sometimes in holes in walls and rocks.

Warblers *(Silviidae)* This is another large family confined to the Old World; many of its members have strong, resonant voices, in spite of a diminutive build. Many of the *Silviidae* are migrants, but the distribution of species does not extend to the coldest parts of the globe. Generally speaking *Silviidae* love to hide in the midst of bushes or in the thickness of reed-beds. Consequently it is much easier to hear them than to see them. Hardly any of them are brightly coloured, but, as said before, they have a more than remarkable voice. The famous Tailor Bird from South East Asia is a member of this family.

Babblers *(Timaliidae)* This is a family with very few common features. It is distributed across the Old World and contains about 280 species. Usually insect-eating, they are typically found in wooded regions rich in undergrowth. Plumage is often bright and in some species the song is melodious. In others, as their family name suggests, these birds are garrulous and noisy rather than tuneful. It includes the group called laughing thrushes, of which the White Crested Jay Thrush is a typical example. Its cackling call reminds one of the Australian Kookaburra. The Pekin Robin is also classed under the babblers, but it is exceptional because of its sweet song.

Vireos *(Vireonidae)* The members of this family are of small size. To it belong about 42 species spread throughout America. They are essentially insect-eating, arboreal birds. Their song is very melodious and comparable to that of the *Silviidae* of the Old World. The song period of some species extends to late summer or early autumn.

Jay

Cordon bleu

Hangnests and troupials *(Icteridae)* This family embraces 49 species which are scattered throughout the New World. Its members have a characteristically yellow plumage, some more so than others. Some of the species have parasitic habits corresponding to the European cuckoo. Many of them are excellent songsters. They vary in size from the caciques as big as crows to the cowbirds and troupials nearer the size of thrushes. Hangnests weave the most beautiful nests with grasses and strips from palm fronds which they suspend from the tips of branches. These are elongated and purse-shaped and the birds enter from below. Hand-reared specimens make charming pets. These yellow and black birds are called orioles in America but are not to be confused with the Old World orioles to which they are unrelated.

Finches *(Fringillidae)* A family of seed-eating birds found throughout the world, even in the coldest and most inhospitable regions. They are usually sedentary. One of their characteristics is the cleverness with which they build their nests. Their song, though not always melodious, is well-developed in many members of the family. Some of the *Fringillidae* are also brightly coloured. They combine two much-prized qualities – a fine song with a beautiful appearance. Typical of this group are the American cardinals and the European goldfinch, bullfinch and linnet.

Sparrows and weavers *(Ploceidae)* In a broad sense these birds really belong to the sparrow family. The common house sparrow is in fact a weaver. In this family for convenience we have also grouped many small African birds which should perhaps form part of a separate family. All members of the *Ploceidae* family are small in size and, generally speaking, are highly gregarious birds. This gregariousness makes them a menace to the cultivation of cereals, for they alight on growing crops in hordes. Many of them build their nests in colonies beautifully woven from grasses and palm leaves often forming large, bulky constructions in the tree-tops. Some of them are nesting-parasites. Many weavers and their African relations are gaily coloured. Their song is not particularly strong.

Reed bunting

Buntings *(Emberizidae)* These are the common buntings that can be found throughout the world, from the arctic regions (Snow bunting) to the tropics (House bunting). In this family are birds which could equally well be classified in the *Fringillidae* family. Typical buntings are ground-living birds, rather gregarious in winter. Their plumage is usually brownish, but some species are brightly coloured in the summer, for example the Black-headed bunting and the Red-headed bunting, of Europe and Asia. Generally speaking their song is neither strong nor musical, but their call-notes are much varied. Mention should be made of the beautiful American buntings which are far more colourful than those from the Old World. The Painted bunting or nonpareil is blue, green and scarlet and the Rainbow bunting is azure blue and yellow.

Mocking birds *(Mimidae)* A family originating in the tropics and now only found in America. Its members have characteristics which are intermediate between those of the thrushes and the wrens, two of their features being a long tail and a slightly curved bill. A few are brightly coloured. Their main feature is the song, very melodious and extremely pleasant to hear. The American mocking bird, for example, sings all the year round and at all hours of the day. Because of this it is very popular in its native country. Its name is derived from its ability to mimic the songs of other birds with great accuracy.

Pittas *(Pittidae)* These beautiful birds are about the size of thrushes. They live mainly on the ground in deep forests and progress by long hops. They have long legs and very short tails. They are brilliantly coloured in scarlet, blue, green and yellow. They come from Africa and Asia. Whilst not exactly song birds their whistling calls are far reaching in the jungle where they sometimes call at night.

Bulbuls *(Pycnonotidae)* There are 114 species of bulbuls. All come from Africa and Asia. About the size of thrushes, they are not brightly coloured although their plumage patterns of black, white and brown, offset in some species with touches of red or yellow, are striking enough. Some species are crested. Many of them are loud if somewhat repetitive songsters. They are active and inquisitive and often frequent gardens and cultivated areas where their fondness for fruit can be a nuisance.

Leafbirds and fairy bluebirds *(Chloropseidae)* These brightly coloured birds about the size of thrushes all come from S.E. Asia. The leafbirds are coloured mainly green with touches of black, yellow and blue. There are 8 different species. These birds are called Fruitsuckers by bird fanciers with whom they are popular as aviary birds. Needless to say their mainly green plumage is a wonderful camouflage among the leaves of the forests where they live.

Fairy bluebirds are much more conspicuous, being black below and bright blue above with red eyes. Leafbirds and fairy bluebirds are good songsters.

Wrens *(Troglodytidae)* America is the home of all but one of the 59 species. The exception is the little brown wren – the Jenny wren of folklore which is distributed in several races throughout Europe and Asia. It also occurs in America where it is called the winter wren.

Not all wrens are as small as this well-known bird. Some, such as the Cactus wren from Mexico, are 8 inches long. All the wrens are good songsters.

The European wren likes to roost in holes and nestboxes put up by man. In very cold weather literally dozens will crowd into one box for warmth at night and it is quite remarkable to watch them popping out one by one in the morning in an apparently never-ending stream.

Tanagers *(Thraupidae)* This family of 222 species all come from the Americas. Varying in size from sparrows to small thrushes they are some of the most brilliantly-coloured birds in the world. Their gaudy hues of blue, scarlet, green and yellow need to be seen to be believed. In many ways they resemble finches and cardinals but whilst a few eat a certain amount of seed their diet is mainly fruit and insects. They are widely distributed throughout tropical South America. A few have penetrated North America.

European robin

Recording the song of birds

Many ornithologists and bird watchers recognise the importance of bird song as an aid to recognition and, with the existence of so many portable tape recorders in domestic use, it is not surprising that many use them as a means of capturing sounds for later identification. It is, of course, a far cry from making recordings for this purpose to making recordings of a professional standard, but it is surprising how many who take to using a recorder in the field become completely absorbed in the subject.

Suitable equipment for making recordings can cost from as little as £20 to over £1,000, and, while recordings made with expensive equipment can be very good, those made with the cheapest can be very acceptable.

Redstart

Cassette machines come within the cheapest range and have the advantage of being light in weight and economical on batteries. Their frequency recording range is usually restricted when compared with reel to reel machines, and the signal to noise ratio is much inferior. The use of low noise, high energy and chromium dioxide tapes help in these respects but not many portable cassette machines are available to take advantage of chromium dioxide tape, and those that do are understandably more expensive. Another feature so very often incorporated in cassette machines is automatic gain control; this is to be avoided because between each burst of song the ambient noise builds up and the effect is most disconcerting. Ironically, automatic gain costs the manufacturers more to incorporate so that for our purpose, unless the machine has both manual and automatic control, the cheaper machines are best. It must be pointed out that because of the limited frequency response, attempts at recording many species are at best bound to be inferior to recordings made with reel to reel machines.

All recordings made in the field must sooner or later be edited to extract wanted parts and to remove extraneous noises. Cassettes cannot be edited in the normal way by using a razor blade and splicing tape, and so the cassette must first be copied on to standard tape. If the would-be recordist already has a mains reel to reel machine this should present no problem. However, if resources are limited it would be more prudent to purchase a reel to reel portable machine, even if it will cost more than a cheap cassette machine. The reel to reel machine will, in any case, produce better recordings.

Reel to reel machines have the disadvantage of being bulkier and heavier and are more expensive on batteries but, in general, such machines are essential if the quality of recording is important. For the very best quality the machine should have a top speed of at least 19 cm/sec. The faster the speed the better the frequency response and signal to noise ratio. Finally, as an added bonus, editing tapes recorded at higher speeds becomes much easier. A great deal more could be written about the choice of a reel to reel machine and many things should be considered. Is the machine easy to operate, in the dark as well as in daylight? Are the carrying arrangements satisfactory and how easy is it to change reels?

After the field recorder, the next item on the list to be considered is the microphone. Cassette machines and cheap reel to reel machines are usually supplied complete with a microphone, but while such a microphone can be used, in general such machines will do justice to a better type of microphone than the one provided.

The better reel to reel machines are usually supplied without microphones, thus leaving the choice open to the recordist from the outset.

The most suitable microphone for the bird song recordist is of the moving coil pattern, nowadays commonly referred to as dynamic. These microphones are robust and are relatively insensitive to handling and wind noise. Dynamic microphones can be obtained with omni- or uni-directional properties and some are available with a bass roll-off characteristic. For our purpose this latter characteristic can be most useful as it helps in reducing low frequency interference from nearby traffic, aircraft, etc. A unidirectional characteristic is also helpful in the same context, although a directional characteristic can be imparted to an omni-directional microphone by mounting it in a reflector. The impedance of the microphone must be less than or equal to the input impedance of the recorder to ensure efficient transfer of the signal.

Zebra finch

For example, many microphones are made in versions having low, medium and high impedances, say 25 ohms, 200 ohms and 50,000 ohms. For a recorder having an input impedance of 600 ohms, the 200 ohm microphone would be most suitable, the 50,000 ohms model would be entirely unsuitable and the 25 ohm not quite so good as the 200 ohm model. Matching transformers can be purchased to overcome this problem but in general it is much better to have a direct connection rather than having to carry an extra piece of equipment which is an encumbrance and can also attenuate what is often already a weak signal. It is perhaps worth mentioning that while most recording is normally carried out with the microphone quite near to the recorder, it may be desirable to extend the leads for some reason. This is perfectly permissible with low and medium impedance microphones, but extending the leads of high impedance microphones without resorting to matching transformers results in a serious loss of the higher frequencies. In making a final choice of microphone ensure that it has an output of at least 0·2 millivolts per microbar because many of the signals encountered are quite small and outputs below this figure will not give a strong enough signal to modulate the tape adequately unless resort is made to a separate microphone preamplifier. There is nothing wrong with using a microphone preamplifier provided it in itself does not produce noise but it is another piece of equipment to carry, requires its own battery and generally should and can be avoided in general recording work. More expensive microphones may have a better overall response but often their output is low and in any case good frequency response at low frequency is seldom required in recording bird song.

The use of a reflector has been referred to previously and this without doubt, providing it is used correctly, is the most important accessory available to the bird song recordist.

Besides being directional, thus enabling the recordist to single out a chosen subject, the device gives a greatly increased signal. The device may be likened to a car headlamp in reverse and while the analogy should go no further it does illustrate the concentration and intensity that occurs.

American white-breasted nuthatch

The lowest frequency that can be reflected is governed by the diameter of the reflector, and it is found that a diameter of not less than 20 inches will reflect nearly all the frequencies encountered in bird song, certainly all those in the range of warblers. A reflector larger than this will extend the lower frequency limit perhaps unnecessarily and give more gain but will be found to be more cumbersome. In general, a reflector of 20 inches will be found adequate and convenient to handle.

Many focal lengths have been tried giving varying angles of acceptance, but from a practical point of view a focal length which just encloses the microphone has the advantage of giving some protection to the microphone from wind. Thus a 20 inch diameter reflector with a 5 inch focal length is five inches deep and meets this requirement.

Pied flycatcher

Using this equipment in the field is largely a matter of experience. More usually the recordist hand-holds his reflector in his right hand and operates the recorder slung on his side with his left hand. This method requires considerable practice and experience to avoid handling noise but enables the recordist to pan his subject and maintain general mobility. Needless to say, the recordist must get as close as possible to his subject and stand very still during actual recording. Alternatively the reflector can be mounted on a tripod but in this position it is not always possible to keep the reflector focussed on the subject and instant mobility is lost.

The use of an open microphone is yet another alternative but here the leads usually must be extended and the microphone placed in position to which it is hoped the subject will return to and sing. This method calls for the utmost patience and a detailed understanding of the subject's behaviour. Few recordists depend entirely on this method.

It will be seen that recording bird song calls for techniques not normally employed in any other form of recording, mainly due to the fact that the subject cannot be rehearsed and is largely unpredictable in its behaviour. The subject, therefore, presents a very satisfactory challenge and a most interesting pastime.

Song birds: habitats and nests

Many birds migrate to the inhospitable regions of the extreme north solely to reproduce their species and then, two or three months later, start the long journey back to more temperate zones or tropical areas of Africa, Asia and America. It is logical that the greatest concentration of birds is found in areas which are more favoured by nature, that is to say in the tropical and temperate belt of the world.

As well as there being a widespread geographical distribution of birds, the habitats of birds are also varied and scattered. There are, for instance, birds which dwell exclusively or almost exclusively in rocky areas, both in the mountains and on the plain and both in the cold north and in deserts. In the quarries and among the rocks they find food, shelter and sites to build their nests. Among the ground birds – so called because they always live on the ground – there are some that are capable of living in the most inhospitable deserts, while others prefer the plains. Of these birds many are excellent songsters, such as skylarks and the Calandra larks. These desert birds manage to reproduce in the most unfavourable conditions where there is an almost complete lack of water and humidity.

But it is in forests that a really impressive number of birds is found, from tropical forests, Alpine woods, hunting groves or scattered woodland; all have a bird population. It is in forests that the highest density in the ornithological population is recorded. The fact that many birds live in forests does not tell us much. What is interesting is the variety of their habitats, the fact that several species or even whole groups of birds have in the course of time become specialists as regards habitat and now dwell almost exclusively in forests or woods of a particular type.

One of these groups is of birds that inhabit coniferous woods, among whom is the crossbill. With this bird specialisation is so strong that only very rarely does one find it away from conifers even during migration. Its beak is shaped in a curious criss-cross fashion which enables it to open pine and fir cones and to extract from them the kernels on which to feed.

In the tropical rain forests where trees reach fantastic heights, there are horizontal stratifications of various groups of birds. There are birds which always inhabit the undergrowth or the ground, others that prefer the thicker foliage of the trees and others that like to live on the treetops.

Another group of birds, particularly common in temperate zones and consequently in Europe and America, inhabits the undergrowth of woodlands, bushy areas, hedgerows and cultivated areas that contain small thickets.

These habitats are rich in food and offer a safe refuge especially with regard to the building of the nest. The number of ways in which nests can be built are many. Exotic birds display exceptionally clever qualities in this respect, witness the Tailor Bird found in South East Asia that joins two large leaves by sewing them together with a thread and then builds its nest inside them. There are some African warblers that build their nests with thorns.

The small *Ploceidae* build very strong nests in comparison to their size. Gathering building materials can take a pair of birds more than a week of continuous coming and going. The weavers in fact get their name from the elaborate and beautiful nests which they weave from grasses and palm fronds. These are globular in shape and are usually suspended from the tips of branches well out of the reach of predators. The entrance to these nests is situated either in the side or underneath and often has a 'spout' up which the bird shoots in flight. When numbers of the nests are built close together they resemble clusters of dried fruit hanging from the branches.

The materials that make up the structure of a nest are infinitely varied. Small twigs and grass blades are often used as the basis of the construction. Padding materials follow – made of feathers or hair or mud, the latter expertly smoothed when the song thrush is at work. Some birds employ the same materials for making the nest, wherever they live. The chaffinch is one of these; it uses moss and cobwebs which make the nest look similar to the bark of the tree on which it is built. Other species show a predilection for the use of heterogeneous materials in the making of the nest and make a varied choice of site. The redstart and the robin seem to be the most imaginative in site selection. There are stories of their nests being constructed in empty tins, old chimney pieces, post-boxes, wheelbarrows, apiaries and on railway carriages.

In many countries bird-lovers make the search for a nest site easier for many birds by providing purpose-made bird boxes or artificial nests. These can take the form of wooden boxes with a hole in the side to simulate a hole in a tree trunk. These will attract tits, wrens, and nuthatches. An open-fronted box with part of the front boarded across to retain the nesting material will suit blackbirds, thrushes, wagtails and flycatchers. Even house martins are catered for these days by the provision of concrete nests which can be fixed under the eaves of a house.

Bibliography

Armstrong, Edward A., *A Study of Bird Song,* Oxford, 1963

Armstrong, Edward A., *The Way Birds Live,* Dover Publications Inc., New York, 1967

Bent, Arthur Cleveland, *Life Histories of North American Wood Warblers,* Dover Publications Inc., New York, 1966

Blachly, Lou, and Jenks, Randolph, *Naming the Birds at a glance,* Knopf, New York, 1963

Burton, John A., *Birds of the Tropics,* Orbis, 1974

Day, J. Wentworth, *British Birds of the Wild Places,* Blandford, 1961

Fisher, J., *Thorburn's Birds,* Ebury Press, 1967

Fisher, Peterson, *The World of Birds,* Macdonald, 1964

Mackworth-Praed, Grant, *Birds of Eastern and Northeastern Africa,* Longmans Green, 1952–55

Risdon, D. H. S., *Cage and Aviary Birds,* Faber, 1967

Robins, Bruun Zim, *Birds of North America,* Golden Press, 1966

Voous, K. H., *Atlas of European Birds,* Nelson, 1961

Yapp, W. B., *Birds and Woods,* Oxford, 1962

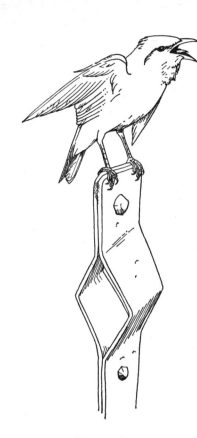

Posture of the Starling when delivering its song

Courtship-flight of the Tree Pipit

Great Reed Warbler singing

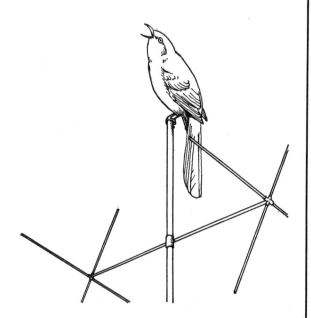

Two birds that make use of TV aerials to deliver their song.
Left: *the Redstart from Europe.* **Right:** *the North American Mockingbird*

Index of birds illustrated

References are to picture numbers
Acrocephalus arundinaceus: see Warbler, great reed
Acrocephalus palustris: see Warbler, marsh
Alaudia arvensis: see Skylark
Anthus pratensis: see Pipit, meadow
Blackbird 55, 56
Blackcap 26, 27
Bluebird 7
Bluethroat 44
Bullfinch 66, 67
Canaries 61, 62, 63
Cardinals
  red-crested 6
  Virginian 5
Carduelis cannabina: see Linnet
Carduelis spinus: see Siskin
Chaffinch 57, 58
Chiffchaff 30, 31
Coccothraustes coccothraustes: see Hawfinch
Cordon bleu 71, 72
Crossbill 16
Erithacus rubecula: see Robin, European
Estrilda melpoda: see Waxbill, orange-cheeked
Fieldfare 48, 49
Finches
  Bengalese 76
  Senegal fire 77
  spice 75
  zebra 73, 74
Fringilla coelebs: see Chaffinch
Galerida cristata: see Lark, crested
Goldfinch, American 1
Hawfinch 68
Hippolais polyglotta: see Warbler, melodious
Lagonistica senegala: see Finch, Senegal

Larks
  crested 8
  wood 14
Leistes militaris: see Meadowlark
Leothrix lutea: see Robin, Pekin
Linnet 64, 65
Loxia curvirostra: see Crossbill
Lullula arborea: see Lark, wood
Luscinia megarhynchos: see Nightingale
Luscinia svecica: see Bluethroat
Meadowlark, red-breasted 2
Monticola saxatilis: see Thrush, rock
Munia punctulata: see Finch, spice
Munia striata: see Finch, Bengalese
Nightingale 41, 42, 43
Oriole, golden 15
Oriolus oriolus: see Oriole, golden
Ouzel, ring 45
Paroaria corunata: see Cardinal, red-crested
Phylloscopus collybita: see Chiffchaff
Phoenicurus ochurus: see Redstart, black
Phoenicurus phoenicurus: see Redstart
Pipit, meadow 13
Pyrrhula pyrrhula: see Bullfinch
Redstarts 32, 33, 34, 35
  black 36
Redwing 46, 47
Rhamphocoelus Brazilius: see Tanager, scarlet
Rhamphocoelus dimidiatus: see Tanager, scarlet-rumped
Richmondena cardinalis: see Cardinal, Virginian
Robins
  American 53, 54
  European 37, 38, 39, 40
  Pekin 10
Serinus canarius: see Canaries
Sialia sialis: see Bluebird

Siskin 59, 60
Skylark 11
Spinus tristis: see Goldfinch, American
Starling 17, 18
Sturnus vulgaris: see Starling
Sylvia atricapilla: see Blackcap
Sylvia borin: see Warbler, garden
Sylvia cantillans: see Warbler, subalpine
Sylvia communis: see Whitethroat
Sylvia melanocephala: see Warbler, Sardinian
Sylvia undata: see Warbler, Dartford
Taeniopygia castanotis: see Finch, zebra
Tanagers
  scarlet-rumped 3
  scarlet 4
Thrushes
  mistle 52
  rock 12
  song 50, 51
Turdus iliacus: see Redwing
Turdus merula: see Blackbird
Turdus migratorius: see Robin, American
Turdus philomelos: see Thrush, song
Turdus pilaris: see Fieldfare
Turdus torquatus: see Ouzel, ring
Turdus viscivorus: see Thrush, mistle
Uraeginthus bengalus: see Cordon bleu
Warblers
  Canada 9
  Dartford 29
  garden 23, 24
  great reed 20
  marsh 19
  melodious 21
  Sardinian 28
  subalpine 22
Waxbill, orange-cheeked 69, 70
Whitethroat 25
Wilsonia Canadensis: see Warbler, Canada

1   **American goldfinch** *(Spinus tristis)*, length 4½ inches (11½ cm). This well-known American bird occurs throughout the United States, breeding in the north and wintering in the south. The male has a seasonal change of plumage. In the breeding season it is like the illustration. In winter it turns olive green like the female but retains the white in the wings. In size it is a bit smaller than the European Goldfinch. It has a pretty song rather like that of the canary. As it is now protected in America it is seldom seen in aviaries. It feeds on seeds of all kinds and prefers shrubby fields and meadows for its habitat.

2 Red-breasted meadow lark *(Leistes militaris)*, length 7½ inches (19 cm). There are several species of Meadow larks, or Marsh birds as they are called in Europe where they are frequently imported as they are popular aviary birds. The Red-breasted species is about the size of a small starling and has rather the same habits particularly when walking about the ground looking for insects. It ranges from Panama and Venezuela south to Brazil, Peru, Argentina and Bolivia. The bird is streaked from above with a scarlet breast. In the breeding season the brown upper parts turn black. It inhabits grassy plains and marshes. It does well in an aviary and thrives on insectile food, seed and insects. Another name for this species is Military Starling.

3 Scarlet-rumped tanager *(Rhamphocoelus dimidiatus)*, length 7½ inches (19 cm). This is a close relative of the Scarlet Tanager and has the same distinctive mark on the lower beak. It occurs in Panama and Venezuela. Its habits and diet are the same as those of the Scarlet Tanager and in captivity it does well on the same diet and treatment, but it is not often offered for sale.

4 Scarlet tanager *(Rhamphocoelus Brazilius)*, length 7½ inches (19 cm). This is a brilliantly coloured member of a very brightly coloured family containing numerous species. Tanagers are a group of fruit-eating birds related to seed-eating finches. In fact some tanagers will eat a certain amount of seed. The hen Scarlet Tanager is reddish-brown with darker wings and tail. In America it is often called the Silver-beaked Tanager because of the prominent white mark on the lower mandible and to distinguish it from the North American Scarlet Tanager, a different species in which the female is yellowish green. The Scarlet Tanager has a short song consisting of a few notes often repeated. It does well in an aviary but is too restless for a cage. Its diet should consist of insectile food, fruit and insects.

5   6

5   Virginian cardinal *(Richmondena cardinalis)*, length 7½ inches (19 cm). This brilliant bird is also quite a good singer – an unusual combination in birds. The hen is duller than her mate, being brown with reddish wings and tail. It has a wide range over the United States and is gradually extending this northwards. It is a popular garden bird in America and a favourite with aviculturists in Europe where it does well as an aviary bird and has often been bred. It builds an open cup-shaped nest in a bush. Its rather heavy bill is well-adapted for cracking seeds but it also eats insects and fruit. In captivity the brilliant plumage fades to a brick red unless the bird is fed properly. Young birds in immature plumage resemble the hen. Its song may be described as a short series of flute-like whistles.

6   Red-crested cardinal *(Paroaria corunata)*, length 7½ inches (19 cm). This fine bird, almost as big as a small thrush, comes from Brazil, Argentina and Bolivia. The sexes are alike but the male tends to depress his crest more than the hen which carries it more erect. The male has a powerful but short song which is quite musical. It builds an open cup-shaped nest in a thick bush. Its food consists of seeds and insects. It does well in aviaries but is too restless to keep in a cage. It thrives on a good seed mixture, plenty of green food and a few mealworms or maggots now and again. Although it comes from warm countries, once acclimatized it is impervious to cold and will live for many years in temperate countries.

7   Bluebird *(Sialia sialis)*, length 5½ inches (14 cm). This is as famous in American folklore as is the Robin in Europe. It is the species about which the popular war-time song was written, although the appearance of this American species over the White Cliffs of Dover would have caused a stir in ornithological circles! The bird is well-known throughout North America. It nests in holes, so wooden nest boxes are put up in gardens to attract it. It is primarily an insect eater but takes berries in the autumn and winter. It has a pretty song which is not very loud. The hen is paler-coloured than the cock and lacks the bright blue on its mate's back. The young are spotted like a thrush. The bird in the illustration is the Eastern Bluebird, but there are several species differing in the amount of blue in the plumage. They are protected in the United States so are impossible to obtain except under licence. In former times, however, Bluebirds were found to be easy to keep in aviaries and they bred freely.

8 Crested lark *(Galerida cristata)*, length 7½ inches (19 cm). This lark inhabits central and southern Europe and parts of Africa and Asia. The grey-brown feathers with less defined dark centres and the more pronounced crest readily distinguish it from the Skylark. It frequents roads where it likes to peck at horse manure and to indulge in dust baths. It also frequents the sea shore. It has a melodious song. It nests on the ground, and occasionally in holes in walls. It lays four or five greyish eggs spotted with brown. The sexes are alike.

9 Canada warbler *(Wilsonia Canadensis)*, length 4¾ inches (12 cm). Our illustration shows a male in breeding plumage carrying the distinctive black necklace which is nearly lost in the autumn and winter when he resembles the hen and immature birds. The bird prefers to live in undergrowth in mixed woodland. It breeds southward from Labrador as far as Georgia and winters in South America. The song is described as rapid and varied. It feeds mainly on insects.

10 Pekin robin *(Leiothrix lutea)*, length 4¾ inches (12 cm).
This bird lives in the Himalayas, Assam and China. In Europe, particularly in the last few years, it has become popular as a cage bird because of its liveliness, its beautiful song and its attractive colouring. Its song is loud and vibrant, of a flute-like or whistle-like character, roughly resembling that of the Blackbird, but uttered more decisively. Maintenance in captivity does not present any difficulties; it must be fed on a diet of insectile mixture, and likes fruit, mealworms and maggots.

11

11 Skylark *(Alauda arvensis)*, length 7 inches (18 cm). The Skylark is widely distributed throughout the temperate regions of Eurasia and North Africa. Its tendencies are migratory, especially in the northern part of the breeding territories, but in the south populations are more likely to be permanently resident as in Italy. The Skylark is a bird of the plains. That is to say, its original environment was that of wide open spaces devoid of trees that can still be found in central Asia. This bird has an unusual and characteristic courtship-flight performed at a great height during which time it delivers its clear, melodious song made up almost entirely of sustained and pleasingly modulated warblings. In display the Skylark takes off from the ground and with wide concentric circles mounts higher and higher singing all the time. Eventually it descends with open wings or flutters down with spread-out tail, finally dropping gently to the ground and ending its song at the very moment of touch-down. This usually happens in the spring, but it is not infrequently heard during other seasons. A ray of sun peering through the grey winter clouds and the skylarks are off, into flight, chasing one another and singing.

12 Rock thrush *(Monticola saxatilis)*, length 7½ inches (19 cm). This beautiful bird is distributed throughout Europe, Asia and North Africa. It is particular in its choice of habitat, living exclusively on sunny south-facing slopes of mountains, among rocks and trees in which it perches. Courtship-flight is particularly fascinating: the male perches on top of a rock and takes off from there in a grand manner, climbing almost vertically, and then descending to land on another rock or raised position only a little distance away from its starting point. At the same time it delivers its song with a rich and melodious voice.

13 Meadow pipit *(Anthus pratensis)*, length 5¾ inches (14.5 cm). The distribution of the Meadow Pipit for breeding purposes extends from Europe to the Urals and Asia Minor, but does not breed in the Mediterranean countries, although it is very common as a bird of passage in the winter when it frequents damp grassland, flooded fields, marshes and bogs. Courtship-song is a sharp note swiftly repeated, ending in a final trill. This song is heard during aerial display, while the call, consisting of a very thin often rapidly-repeated note, is heard at all times.

14 Woodlark *(Lullula arborea)*, length 6 inches (15 cm). The Woodlark, as its Latin name 'arborea' indicates, has a habit of perching freely in trees or on high posts in open ground. The name 'lullula' comes from the song, in which the dominant phrase is a sort of lulululululu. The song itself, though lacking the vehemence of that of the Skylark, is far more melodious and consists of a sequence of short phrases fairly varied in composition. Courtship-flight is like that of the Skylark, though the song is not infrequently delivered from a perch in a tree.

15 Golden oriole *(Oriolus oriolus)*, length 9½ inches (24 cm). Breeding grounds of the Golden Oriole are distributed throughout continental Europe and Siberia as far as Turkestan and India. Essentially migratory, the birds arrive at their breeding grounds in May and leave in September. They prefer broad-leaved woodland regions with tall, shady trees in the neighbourhood of water. They do not like coniferous forests. The song is quite unmistakable, being extremely loud and made up of a mellow flute-like phrase. These birds are rarely offered for sale in Britain as they are not easy to breed in captivity. They are, however, available in Mediterranean countries like Italy and Spain.

16 Crossbill *(Loxia curvirostra)*, length 6½ inches (16.5 cm). This bird has a song that resembles the Greenfinch. It is usually interspersed with its call, a sharp 'chup, chup' delivered on the wing and from belts of coniferous trees which it prefers to deciduous. On the ground it is clumsy, and prefers not to descend until forced to do so in order to drink. Its diet consists of the seeds of cones from conifers, berries of various types and some insects. It breeds throughout Europe and northern Asia. It has an affinity with parrots, both in its postures and its cross mandibles. The adult male has plumage of a distinct brick-red, the female is yellow-green and the young greenish-grey.

17–18 Starling *(Sturnus vulgaris)*, length 8½ inches (21.5 cm). The breeding area is Europe, Turkestan, Iran, Siberia and Mongolia. The pattern of migration is particularly complex, as each population has its own special wintering zone. For example, the Starlings of southern Finland winter in France, those of Poland in Spain and North Africa, and those of Hungary in Italy and Tunisia. Its song is not particularly melodious. During the mating season the male takes up a prominent position and frantically agitates its wings, while uttering a continuous chirp interrupted by trills, whistles and flute-like notes. The male is particularly good at mimicking other birds, the song of the Oriole, for example, being one of its specialties. The Starling is not particular about its habitat, being equally at home in town or country and on wild or cultivated ground. For nesting in the breeding season, however, it prefers human habitation or woodland. Characteristic of the bird is its jerky walk and its active curiosity.

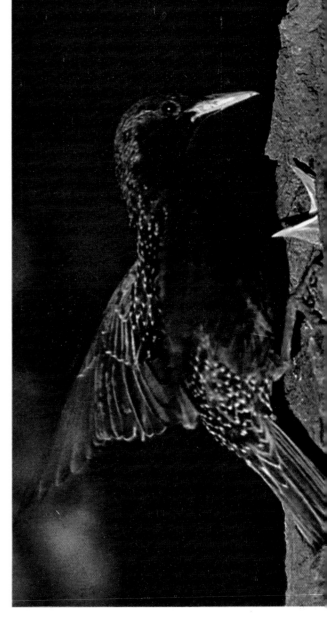

19 Marsh warbler *(Acrocephalus palustris)*, length 5 inches (13 cm). Breeds in Britain, in central and eastern Europe and in Russia as far as the Caspian Sea. In winter it migrates to Africa and Asia Minor. Marsh Warblers are not gregarious birds, yet where the environment is right they will gather in a kind of colony for nesting purposes, and then they give out a concert of songs, aggressively delivered and in which one can distinguish trills, whistles and imitations of other birds. This display goes on incessantly, even throughout the night and especially during full moon. In Italy the Marsh Warbler is often visited by the Cuckoo who deposits an egg in its nest.

20 Great reed warbler *(Acrocephalus arundinaceus)*, length 7½ inches (19 cm). This bird, which is the largest of the European warblers, nests in Eurasia and North Africa. In winter it migrates to tropical Africa, Indonesia and Malaysia. Its habitat is limited to the marsh in which it spends all its time. In spring and summer the voice of this bird can be heard at a great distance. The male delivers its song usually while clinging to a reed. Song is made up of a few repetitive drawling notes, which somewhat resemble the croaking of a frog although better modulated and pleasing to the ear. It often sings at night.

21 Melodious warbler *(Hippolais polyglotta)*, length 5 inches (13 cm). This species nests only in France, Spain, Italy and North Africa and in winter it migrates to tropical Africa. It is a member of the genus Hippolais, a group of birds very much alike that appear to have divided the Mediterranean basin into distinct areas, within which each species nests without interference from the others. Courtship-song of the male is very complex, consisting of repeated staccato notes which often include a series of hurried variations sometimes accompanied by mimetic sounds. The male delivers its song from the top of a tree, where its yellow breast becomes most conspicuous.

22 Subalpine warbler *(Sylvia cantillans)*, length 4¾ inches (12 cm). This bird, like many Silviidae, is a typical inhabitant of the Mediterranean scrubland, a complex mantle of vegetation that covers, or used to cover, all the countries along the Mediterranean sea. The Subalpine Warbler breeding in Italy can be found up to reasonable altitudes on the ridge of the Appenines. In winter it migrates to tropical Africa. Its song is similar to that of the Whitethroat, but the notes are more prolonged and the phrase more sustained, though rather muddled at times. Due both to its small size and type of habitat it is not an easy bird to observe.

23–24 Garden warbler *(Sylvia borin)*, length 5½ inches (14 cm). Though the plumage of this species is not very conspicuous, the Garden Warbler is, unfortunately, shot for food in Mediterranean countries. In autumn, migration time, this bird is plump due to the abundance of fruit on which it feeds, especially figs. In fact its Italian name of Beccafico means 'fig eater'. The Garden Warbler nests in Europe and in Siberia as far as the river Yenisei, and in winter migrates to tropical Africa. Its song is remarkably melodious, more sustained than that of the Blackcap and more evenly flowing.

25 Whitethroat *(Sylvia communis)*, length 5½ inches (14 cm). The Whitethroat nests in Europe, Turkestan, Asia Minor and Siberia as far as the Yenisei. In winter it migrates to tropical Africa. This bird is a typical inhabitant of hedgerows, liking rough ground with tangled vegetation, scrub and undergrowth in general, and even that of cultivated areas – though it cannot be found near to human habitation as can the Blackcap. Its song is made up of short bursts of a rapidly uttered warble less melodious, less rich and also less powerful than that of the Blackcap. This species, like the Marsh Warbler and the Great Reed Warbler, is also one of the Cuckoo's favourite victims, so much so that in Germany it has come to be called Kuckucksamme, which means Cuckoo's nurse.

26–27 Blackcap *(Sylvia atricapilla)*, length 5½ inches (14 cm). The Blackcap breeds in Europe, Eastern Siberia, Asia Minor, North Africa, the Canary Islands and the Azores. Winter migration takes it to Africa and the southernmost countries of the Mediterranean basin. The Blackcap is one of the commonest and better known of the Warblers. Its favourite habitats are open woodland areas, copses, parks and gardens with plenty of undergrowth, bushy places, shrubberies and brambles. The fame of the Blackcap is rightly due to its song, which consists of a long phrase of clear, vibrant rising notes sometimes accompanied by harsh drawling sounds or a great burst of rapidly uttered notes.

28 Sardinian warbler *(Sylvia melanocephala)*, length 5¼ inches (13.5 cm). The Sardinian Warbler is another inhabitant of the Mediterranean scrubland and is, in fact, largely resident there, although many of the species will winter in North Africa and Arabia. The Sardinian Warbler is often heard but seldom seen, because it prefers to conceal itself in the middle of thick dense evergreen vegetation or tangled foliage where access is difficult. Courtship-flight, carried out in the spring, is similar to that of the Whitethroat and, in some ways, so is the song although the phrase is much more prolonged and melodious.

29

29 Dartford warbler *(Sylvia undata)*, length 5 inches (13 cm). This species nests only in Spain, France, central and southern Italy and southern England, though in England breeding is often restricted by climate to local areas. The Dartford Warbler likes to inhabit bushy areas with thick undergrowth and is usually found on heaths, commons and scrubland. Its song is heard in the spring and is made up of harsh, often muddled notes, mingled with clear ones. It delivers its song while carrying out a typical courtship-flight.

30–31 Chiffchaff *(Phylloscopos collybita)*, length $4\frac{1}{4}$ inches (11 cm). This bird is perhaps the most common of a series of small and insignificant birds widely distributed, from the territorial point of view, throughout Eurasia. Members of the genus Phylloscopus, in spite of their apparent fragility, are known to migrate from the arctic regions to the tropics, sometimes covering a distance of well over 10,000 km. This is no small achievement for a bird that weighs only 8 gr. It is found in Europe, the Caucasus and as far away as central Siberia. In winter the majority migrate to India and Africa, although some can, quite often, be found wintering in Italy. The song of the male consists of two or three alternate notes uttered in an irregular sequence, which makes it easily recognizable. Though in summer the Chiffchaff is strictly a bird of forest and woodland, in winter it frequents open spaces and cultivated areas where bird watchers can often hear its characteristic call.

30

31

32–33–34–35 Redstart *(Phoenicurus phoenicurus)*, length 5½ inches (14 cm). The Redstart is distributed throughout Europe, Asia Minor and central Siberia. In winter it migrates south as far as Africa and Arabia. Although not particularly exacting in its choice of habitat, the Redstart prefers dry hilly or low mountainous areas with plenty of trees. In spring the male is noticeable for its beautiful colours, its liveliness and the character of its song. The latter is always delivered from a perch situated in a prominent position, as, for example, a television aerial. Song is made up of short melodious phrases of one long and two short notes, and a warbled trill.

36 Black redstart *(Phoenicurus ochruros)*, length 5½ inches (14 cm). Breeds in Europe and Asia as far as Mongolia and the Himalayas. In winter it migrates to tropical Africa, India and Malaysia. This bird could be described as the Alpine variety of the common Redstart; it breeds at a higher altitude in mountains. Its song is less melodious than that of the Redstart, with more strident and hurried notes.

37–38–39–40 European Robin *(Erithacus rubecula)*, length 5½ inches (14 cm). Breeds in Europe, Siberia, Asia Minor, the Canary Islands and the Azores. In winter it migrates southward. The Robin is an extraordinary bird, mainly because of its familiarity with man and for its natural habits. In spite of its small size it is extremely belligerent and will not tolerate any competitors inside the territory vital for the maintenance of its family. This causes it to give voice very frequently to its song, which consists of short varied phrases, some liquid in quality, some shrill, sometimes even strident in tone. The phrase is, on the whole, well constructed. The Robin is famed in popular folklore, and there are many legends concerning the origin of its red breast.

41-42-43 Nightingale *(Luscinia megarhynchos)*, length 6½ inches (16.5 cm). Inhabits Europe as far as the Ukraine, Asia Minor and North Africa. In winter it migrates to the tropical regions of Africa. The Nightingale likes to live in wooded areas or in tangled and bushy places where it can hide. It is a ground-feeder and lives mainly on a diet of small insects. Its nest is nearly always built on the ground. The song of the Nightingale consists of long varied phrases, made up of a succession of repetitive notes and final variation, and is remarkable for the beauty of its melody, richness and vigour. The literature on this bird is impressive; even poets with no specialised knowledge of birds give lengthy descriptions of the beauty and quality of its song; in fact when writers wish to describe anyone with a remarkable voice they compare her to a Nightingale. Similarly the common names given to such birds as the River Nightingale and the Japanese Nightingale reflect the beauty of their song. The Nightingale loves to sing at dawn and at sunset, and during the night. It also sings in the daytime but the song is less obvious in competition with that of other birds.

44 Bluethroat *(Luscinia svecica)*, length 5½ inches (14 cm). Nests in northern Eurasia, in Turkestan and Manchuria. In winter it migrates to Africa, India and southern Asia. The Bluethroat likes to live in the tundra, in regions of undergrowth and bushes, and especially in damp and marshy environments near the edge of forests. Song consists of metallic-sounding staccato notes followed by some precipitate final notes. On the whole the song is musical and varied, and similar to that of the Nightingale. In Lapland this bird is called 'Singer of the hundred voices'. The Bluethroat usually delivers its song from the top of a bush, but will do so also during its courtship-flight.

45 Ring ouzel *(Turdus torquatus)*, length 14 inches (28 cm). The Ring Ouzel is one of the few birds known to have originated in the Alps. It subsequently spread to other European mountain groups and the subarctic regions. This strong bird takes the place of the Blackbird in environments which are alpine, and can be found up to the limit of arboreal vegetation in the highest belt of rhododendron bushes, pine trees and larches. Its nest is bulky and well-made. It feeds on insects, and berries in the autumn. In winter it migrates southwards as far as North Africa, often mixing with Song Thrushes and Blackbirds. Its nature is as wild as that of the Mistle Thrush. Its song consists of short phrases made up of a few rather harsh notes, less pure in quality than those of the Blackbird.

46–47 Redwing *(Turdus iliacus)*, length 8¼ inches (21 cm). Inhabits the northernmost part of Europe and Siberia. In winter it migrates to the south as far as North Africa. In Italy it appears regularly and in large numbers in the autumn and stays until March. This thrush is very similar to the Song Thrush, but can readily be distinguished by its buff eyebrow stripe and reddish flanks from which it gets its name. It lives in wooded areas to the edge of the tundra and in coniferous and birchwood forests. The song is a short sequence of repeated flute-like notes. Its call, 'seeih, seeih' is also uttered in flight.

48–49 Fieldfare *(Turdus pilaris)*, length 10 inches (25.5 cm). Breeds in central and eastern Europe, Russia and Siberia as far as the Yenisei. In winter it migrates further south and some fly to reach North Africa. It is a regular winter visitor in Britain. The song of the Fieldfare is a long and varied musical phrase, somewhat feeble and with some harsh and squeaky notes. Its call, heard at a great distance and uttered also in flight, is a harsh 'cha cha cha chak, cha cha cha chak'.

50–51 Song thrush *(Turdus philomelos)*, length 9 inches (23 cm). Inhabits Europe, Asia Minor and central Siberia. In winter it migrates to the south as far as North Africa and is quite common in southern Europe. The Song Thrush nests in the Alps and the Appenines even at low altitudes. The male delivers its song in the spring from the top of a tree, a melodious, clear and vigorous song, consisting of a series of rich and varied phrases made up of pure flute-like and whistled notes. Many people prefer its song to that of the Nightingale for sheer beauty.

52 Mistle thrush *(Turdus viscivorus)*, length 10½ inches (27 cm). This large thrush, resident throughout Europe, is less migratory than other birds of the same genus. Its scientific name shows its curious partiality for the milky berries of the mistletoe. The Mistle Thrush likes to inhabit wooded areas on plains, hills and mountains. Its song is similar to that of the Blackbird, but far less varied, with simple repetitive phrases; its voice, however, is vigorous, flute-like and stronger than that of the Song Thrush. It is one of the earliest singers of the year and can often be heard in February.

53–54   American robin *(Turdus migratorius)*, length 8½ inches (22 cm). This is a very different bird from the European Robin of folklore, although it is equally well-known in North America where it is a familiar and popular garden bird. In size and habits it resembles the European Thrush and Blackbird. It builds a similar open cup-shaped nest lined with mud. It feeds on insects and fruit. It has a fine song. It is now a protected species in the United States so is almost unobtainable as an aviary bird. In the old days when specimens were obtainable it proved an excellent aviary subject and bred freely. The juvenile bird has a spotted breast until its first moult into adult plumage.

55–56   Blackbird *(Turdus merula)*, length 10 inches (25 cm). This is a very common bird throughout Europe. It occurs everywhere, even in the middle of towns. It prefers to inhabit areas with plenty of trees, it likes the undergrowth and cultivated areas surrounded by hedgerows and bushes inside which it hides. Essentially insectivorous it feeds also on wild berries, cherries, grapes and figs. The Blackbird nests early in the season, often in March when the countryside is still bare. At this time it builds its nest in stacks of wood, on tree stumps and even in sheds. The song of the Blackbird is well-known. It has a flute-like quality and is made up of musical phrases, variously modulated.

53

54

55

57-58 Chaffinch *(Fringilla coelebs)*, length 6 inches (15 cm). Nests throughout Europe and as far east as northern Siberia, and in North Africa, the Azores and the Canary Islands. Although usually of sedentary habits, the Chaffinch is migratory in the northern part of its breeding territories. The song of the male is a well-known musical phrase, loud and vibrant, repeated at regular intervals. One can single out many variations in the number of notes which make up the phrase. The Chaffinch is often referred to in popular folklore, and our saying 'A bird in hand is worth two in the bush' in Italy becomes 'Better a Finch in hand than a Thrush in the bush'.

59–60 Siskin *(Carduelis spinus)*, length 4¾ inches (12 cm). This small finch nests throughout Europe and as far east as western Siberia. It does not generally breed in the Mediterranean regions and if found there it is in the neighbourhood of the Alps. In the winter Siskins gather in large flocks and migrate south. It is then that one finds them in large numbers in southern Europe, when unfortunately they are easily netted due to their tameness and their familiarity with people. Both these characteristics make them popular as cage birds, where they distinguish themselves for their liveliness and their intelligence. They are often capable of learning a whole series of tricks such as pulling up, with their beak, a small bucket from a small well. The Siskin's song consists of a medley of mixed notes, some a little harsh, but on the whole quite pleasant to hear.

61–62–63 Canary *(Serinus canarius)*, length 5 inches (12.5 cm). This species is confined to the Canary Islands, the Azores and Madeira where it inhabits woodland regions, especially of coniferous trees. It breeds also in gardens, orchards, cultivated areas and vineyards, and up to a certain altitude on hills and mountains. Courtship-flight is similar to that of the Serin, consisting of a series of rapid flights followed by slow, gliding, almost bat-like movements. Its song is a series of varied and extremely melodious crescendo trills. The Canary was first imported by the Spaniards and soon became widespread in England, France, Holland and Germany due to the ease with which it reproduces in captivity. This fact has led to the variations and mutations which are typical of many types of domestic animals. Canaries can be divided into three groups. The first group is the Song Canaries, usually yellow in colour, green, orange and creamy white. Well-known varieties of this group are: the Harz Mountain Roller, the Malinois, the Timbrado (Spain) and the Columbus Canary (North America). Amongst these the best and better known is the Harz Mountain Roller of German origin. The second group comprises Canaries which are admired for their shape and their perching habits; it includes variations of a rather large size (up to 22 cm in length). It is divided into English varieties (Yorkshire, Norwich, Lizard, Border, Lancashire, Crested, Gloucester and Frilled); varieties from northern and southern Holland, Parisian varieties; Gibber Italicus and Belgian varieties. In the third group there are coloured varieties derived mainly from the Roller and the Border varieties and which include a whole series of colour tones among which deep orange is particularly beautiful: this has been achieved by cross-breeding with the Venezuelan Hooded Siskin, a bright red little bird with a black head, wings and tail.

62　63

64–65 **Linnet** *(Carduelis cannabina)*, length 5¼ inches (13.5 cm). Nests in Europe, Russia, Turkestan, Asia Minor and North Africa. Those from the north migrate south as far as North Africa, and as a bird of passage it is common in southern Europe. The Linnet is one of the few *Fringillidae* which completely changes its colour from spring to autumn. In the spring the male has a crimson breast and forehead; in the autumn it becomes streaked and brownish. Its song is melodious, being made up of a series of rapid and varied musical twitters. As a cage bird it is very popular, being easy to maintain in captivity and breeding freely in aviaries. It crosses readily with the Canary producing hybrids which sing very well.

66-67 Bullfinch *(Pyrrhula pyrrhula)*, length 5¾ to 6¼ inches (14.5 to 16 cm). This graceful finch has a wide territorial distribution, nesting in the Azores, throughout Europe, Siberia and Japan. It is of arboreal habits and is also sedentary, though not in winter when it is an irregular migrant in the northern territories. Its song consists of a phrase made up of varied notes, not very loud and slightly shrill. It is very popular as a cage bird for the beauty of its plumage. Must be fed on seed, fruit tree buds, and the berries of elder, hawthorn and blackberry as they come in season.

68 **Hawfinch** *(Coccothraustes coccothraustes)*, length 6½ inches (16.5 cm). This bird gives the impression of being bull-necked and is equipped with an enormous bill. Its usual note is a sharp, clicking 'tzik-tzik' delivered from a perch on the topmost branches of tall trees, or in flight. Its song is heard from mid-March until May. The Hawfinch breeds over most of Europe. It prefers to build its nest in fruit trees, but has also been known to nest in thorn bushes and forest trees. Its bill is used for splitting fruit stones, hence its liking for orchards, and it is sometimes a nuisance to farmers because of its habit of shredding pea-pods.

69

69–70 Orange-cheeked waxbill *(Estrilda melpoda)*, length 4 inches (10 cm). Nests in west-central tropical Africa south of the Sahara. It is a sedentary bird, confined entirely to those regions where it frequents grassland and cultivated areas. It is gregarious all the year round. It builds a dome-shaped nest with a side entrance. In captivity it breeds well and may produce several broods in a season if given some insect food. It does well if fed on a diet of canary seed, millet, greenfood, and seeding grasses, and must be given plenty of water to bathe in. Its song is feeble and not remarkable in any way. The bird gets its name from the patch of orange each side of the head; the beak is light red. Plumage is brown on the back and grey on the breast. The long black tail is wagged from side to side when the bird is excited. In recent years the Orange-cheeked waxbill – which can do damage to seed crops – has been introduced into the West Indies from Africa, and has colonized the Puerto Rican canefields.

70

71

71–72 Cordon bleu *(Uraeginthus bengalus)*, length 4¼ inches (11.5 cm). Nests in tropical Africa south of the Sahara, and is sedentary and confined entirely to those regions. This bird is one of the best representatives of a large group of *Ploceidae* which live in the savannah and in cultivated areas where they feed essentially on seeds and insects. It is very common near the villages in undeveloped areas. Its nest is built in the shape of a dome and is often sited in curious places, for example, near a wasps' nest. It sometimes uses nests which have been abandoned by other birds. Partly due to its small size, the song of the Cordon Bleu is rather feeble and consists of three notes of which the last is longer than the previous ones. Occasionally it utters a kind of squeaking medley. It is popular as a cage or aviary bird where maintenance is easy – though it must be well protected from the cold. Diet consists mainly of millet, greenfood, seeding grasses and fruit. It reproduces easily in an aviary.

72

73 74

73–74 Zebra finch *(Taeniopygia castanotis)*, length 4¼ inches (11 cm). Is native to Australia and does not migrate. It inhabits cultivated and bushy areas, feeding on grain. It nests usually in cavities in trees and buildings. The nest is bulky and made of grass. In captivity the Zebra Finch rivals the Canary and the Striated Finch as a bird easy to breed in aviaries. As such it has several mutations which have been developed into quite a few different colour varieties. Because of its popularity as an aviary bird it is now bred extensively in captivity, particularly in America and Britain, and in a sense has become a domesticated bird of these countries of adoption. The colouring in the wild is grey, with a white belly, chestnut cheek patches and a band of chestnut down each flank, spotted with white. The beak is red and the legs pink. The chief colour mutation that has occurred has resulted in a silvery grey variety. Other mutations have given us both fawn and white varieties. The latter is pure white with no markings. All retain the red beak.

75 Spice finch *(Munia punctulata)*, length 4¾ inches (12 cm). Breeds in India, Indochina, southern China, the Philippines and Indonesia. It is sedentary and lives in groups in open cultivated areas. The genus *Munia* comprises numerous species of birds of similar living habits that feed mainly on seeds. In its natural environment it is a common and friendly bird. It does well in captivity, can reproduce and will cross-breed with the Marsh Harrier and the Silver Bill.

76 Bengalese finch *(Munia striata)*, length 4½ inches (11.5 cm). This bird is really a domesticated variant of a wild species called the Striated Finch, but at one time there were doubts as to the truth of this because the domestication took place so long ago, possibly in Japan. The Striated Finch breeds, however, in India, Indochina, southern China and Malaysia. It is of sedentary habits and feeds on seed. Its song is a whispered warble, or trill, rather feeble in tone – not surprising considered the small size of the bird. The Bengalese Finch, like the Canary, is an artificial bird, of which many different-coloured varieties are known, such as chocolate and white, fawn and white, pure white, and there is even a crested variety. In Europe and America it is very popular as a cage bird because of the remarkable ease with which it reproduces even in a cage.

77 Senegal firefinch *(Lagonostica senegala)*, length 3¾ inches (10 cm). Confined to tropical Africa south of the Sahara, it is usually gregarious all the year round, living in large groups and frequenting areas in the vicinity of villages. It prefers places with plenty of trees and bushes in which to hide. The nest of the Senegal Firefinch is built in a natural cavity of some sort. In captivity the bird feeds essentially on millet. Breeding gives uncertain results due to the attention which must be given to the rearing of the newly-hatched young that require small insects.